Cora A. Morse

Yosemite As I saw It

Cora A. Morse

Yosemite As I saw It

ISBN/EAN: 9783742810144

Printed in Europe, USA, Canada, Australia, Japan

Cover: Foto ©berggeist007 / pixelio.de

More available books at **www.hansebooks.com**

Yosemite as I Saw It ❧ ❧ ❧ ❧

DR. CORA A. MORSE,

SAN FRANCISCO

"Take furrows upturned by the ploughshare of God.
The mountains which froze where the billows had trod.
The huge of the elements over them swung,
In the wild chimes of Nature there rhythm was rung."

Goethe

SECOND EDITION

DEDICATION

To my friend,

Mrs. Sarah M. Kingsley.

My ideal of perfect womanhood, the strength and inspiration of my life, with the love of my heart, and the hope of her sustaining power to the end of my service to the world, I gratefully inscribe this book.

PREFACE.

THESE letters, written at intervals during my stay in Yosemite, to husband and friends who remained at home, were not intended for the public.

At the earnest solicitation of many friends, however, I have placed them in more permanent form.

In so doing, I lay no claim to literary merit and have found no time from my busy life to correct any discrepancies which may appear. They were written under the inspiration of nature's revelations, and from a heart too full to contain its joy unexpressed.

With no further apology I submit them to my friends and to the reader.

THE AUTHOR.

ONLY a wave from the ocean of thought,
A shell from the beach of life.
A word of the message scarcely caught
During rest from worldly strife;
Only a pebble dropped in the sea
Of daily human endeavor,
Just a hint of the mystery
Surrounding our souls forever.

ONLY a glimpse of the artist in me,
One note from the song in my heart,
A spark of light from my Deity,
Of which all nature is part;
Such, dear friends, is this little book,
And with it goes the prayer
That reading, someday, in a quiet nook,
You may sense my presence there.

ON THE WAY.

LEAVING behind us the smoky city, the heated noisy train, and the first wearisome half day's staging from Raymond, let us try if possible to paint a word picture of the visions focused upon the retina, and burned within the brain in living, pulsing pictures never to be obliterated. We find the route much like that of other mountainous regions of California until we near Wawona, the Indian name for big trees, our resting place for the night. Rapidly the panorama changes, the mountains rise higher, the chasms sink deeper and it is a great stretch of imagination that anything Yosemite may have in store is greater or grander. Our spirits rise with the mountains, until the clouds seem reached, and the stars seem kissed in reality. The eye sweeps everything in, from earth to sky; from the towering trees above, to the tar weed with its white blossoms below. The fragrance of the tar weed, not unlike that of witch hazel, together with that of the wild azelia, is lading the air. Nature gently weeps out her tears here and there, over the stony sides of her grand old hills, like a mother, seeking to melt the stony heart of a wayward child; and like the child, these hills adhere to the pattern in their own souls, allowing the mother to weep on, while they carry out the purpose the creator intended. Here are some wonderful formations of burned and broken tree limbs, looking for the world as though the storm and flame had hewn and burned according to design, these wrecks of their undoing. Lying against rocks, and strewn along against trees and brush, is a network of white, coral shaped, skeleton branches, that seem to reach out like the pleading arms of the lost yesterdays, teaching us the same lesson of faded glory and departed opportunity.

Strangely like the human spine and nervous system, are many of the trees, with branches fine and threadlike ramifying everywhere. The roots of these same trees, which have been torn up by storms, resemble the great ganglions of nerves in the body, making it easy to clothe them with intelligence and even wisdom. Many of the tree branches, are covered with a growth of moss, light green in color, shining in the glint of the sun in seven splendid shades of green, as distinct as the seven shades in the diamond opal. Hanging to the limb of one tree, is a grotesque and beautiful growth, partly overrun with vines, and this is called the "Hanging Basket." Another tree is called the "Pitchfork," because of its growing in that shape. Then there is the "Bologna Sausage" tree, hanging full of short stubby limbs in sections, not unlike bologna sausage.

We are now rattling and rolling along, ominously near the edges of chasms, which grow deeper at every turn. The brain grows dizzy at an upward glance and the heart sinks, as the eyes are cast down, when at last we reach the summit. Our first experience in staging down the mountains, almost takes the breath away, until we become somewhat accustomed to the sweep and curve of the roads. As we near the valley, the scene is enhanced by a running stream, clear as crystal, showing the fish at the bottom, and here are hundreds of campers enjoying small fry and the mountain air. Shimmering through the trees is the dying sunlight, touching the leaves with silver and gold, and transforming the dusty road into a street of crimson light, as we wind in and out among the rocks and trees, beside the river, as it rushes onward to the sea.

We can imagine that the New Jerusalem lies not far away, as we revel in the changing light, and listen to the birds singing lullabys to the little ones in their keeping, as they rock them in the wind-swayed boughs, by the flickering light of the all mother's candle. The woods begin to darken with the lengthening shadows, as we near the floor of the valley, when Presto! Change! we sweep through the gates of the long mountain ranges, into the dazzling blaze of a sunset on the mountains. We spring to our feet and wave our hands, while we sing "Hallelujah," as the doors of this heavenly city are thrown open to receive us. In the warmth and glow and grandeur of this new old scene, how the soul rejoices that "I and my father are one." That the light flashing hitherward from the heavens is the same light that flashed thitherward from the misty eyes upturned to meet it: that the "All hail!" of recognition is mutual. Immediately before our enraptured vision, lies against the sky, the peaks of two lofty mountains. These with the space between them, form an undulation not unlike the great waves of the sea. Behind and through them, the sunlight plays, like smiles around the dimpled mouth of a child. A misty hazy blue tinges the whole mountain range. It is darkness in the valley. Along these two majestic peaks the lines of gold dance like fireflies in the night, and just above them a sheet of gold spreads across the sky like an unbroken placid bosom of water. This melts away into the flame and fire of the upper sky. White over all hang some snow white clouds, pink at the lower edges and silver lined above. Slowly this vivid picture fades into yellow crimson and green. Looking "over yonder what do we see?" "Tis the blushing response of the clouds on the further side of the valley. Across this awful chasm "souls clasp hands." The law of affinity draws the sleepy clouds, these countless miles away into the embrace of their lover, the sun, for his good night kiss, which is manifested in the crimson impress he leaves upon their snowy faces. Like so many coy maidens, they seem to play hide and seek with the hills and the valley below.

Night wears on apace, and we sink at last to sleep in the arms of the Infinite, knowing that life abundant, life everywhere reigns, and that "joy cometh in the morning." For stretching away like guiding arms are the higher peaks, that shout and sing of Yosemite; and the dying sunlight is the prophecy that over yonder lies the promised land of greater interest and rarer beauty.

YOSEMITE VALLEY FROM INSPIRATION POINT

STILL ONWARD.

MORNING in the mountains! Morning in the valley! Straight above us all is blue as It than skies. Not a cloud is floating there. But along the Eastern peaks, the faintest tinge of light appears, like a far away reflection from the bosom of some fiery lake. And gradually sifting through into the valley below comes a warmth and glow, such as bottled up sunlight might throw into a darkened room, uncertain at first, then spreading, making luminous the fog around us, and turning our night into day. Like a torrent of water, forced up an inclined plane until it reaches the summit and gushes over, the glory of the sun-line, rolling up from behind the mountains, and creeping over into the valley, impresses the eye, and carries the comparison to the brain.

As the ear would cognize the sound of rushing waters before they broke over the mountain tops to stream down their sides, so the eye cognizes the billows of light, dancing up the mountain side from beyond, and rolling over the field of blue, until visible in currents of light, running over the top, then to the foot of the mountains before and below. So much for the Eastern view of sunrise in the mountains.

Along the Southern slopes, the light vibrates so intensely as to give one the fancy of phantom footsteps. While to the North, one seems to see the spectre's dance, as old Sol spreads on his colors more and more generously, and rolls the mists in splendor, from the murky Western hills, the latest retreat for the goblins and fairies, who now "rise and shine" in the strength of Father God, until his day's reign is over, when they sink again to Mother God's bosom, to while away the hours of night. A new promise fills our being, and new patience possesses the soul, as we realize what a heaven 'twill be to us all, when life's mists are cleared away and life's goblins and fairies are swallowed up in the light of our great wisdom.

The stage awaits us now, to bear us onward to the haunts of the Indian, who discovered and named Yosemite "Great Grizzly Bear." Leaving Wawona we pass the camp where Uncle Sam's troops are stationed, and just beyond, we cross the "Fish River" again. At the bridge, the water gurgles over great smooth

ROYAL ARCHES.

boulders, polishing them like so much crockery. We drink from a spring at this point delicious mountain water, which seems both food and drink, so completely does it quench our thirst. The morning air is a veritable breath from heaven, as we inhale it into our innermost being.

Once across the river, we are in the edge of Yosemite Park. Soon the trees show taller, larger growth, the mountains tower still farther heavenward, the chasms grow darker and deeper, the curves more abrupt, the road narrower, the ascent steeper; and as the horses toil upward we have ample time to view the landscape o'er.

At the bottom of these vast chasms, the river winds like a serpentine thread of light. Great project ing boulders add beauty to the tall trees as we look down upon them. Vines and lupine, together with azelias and dogwood blossoms, relieve the otherwise sombre scene. The sky above is lost now, excepting a far tches here and there, being hid by the lofty pines around us, as they clap their hands and wave us a far ests greeting. The yellow and sugar pines, together with a good sprinkling of fir trees, fill the air with a "Balm of Gilead," soothing to wearied bodies.

Twelve miles of our road lies along these scenes, when suddenly and without warning we swing around on "Inspiration Point," and face the majestic rocks, scarred and serried, wrinkled and gray, cracked and torn, and gaze six miles down the Valley of Yosemite.

On our right is "Bridal Veil Fall," on the left the "Falls of Yosemite," while in front of us, the "Widow's Tears" trickle down the rocky side of the mountains. These mountains are not earth, trees, and projecting boulders, but solid rock from top to bottom, three to ten thousand feet above the sea level. Here and there clumps of trees grow upon them, and everywhere the marks are visible where the waters flow over them as the snow melts in the springtime.

There is no language, no paint that can paint this mighty picture : one feels here the touch of the mas ter mind of the Universe, the tongue clings to the roof of the mouth, the head bends in dumb recognition of the power, which called order out of chaos, caused the land and waters to divide, and the mountains to appear. The magnitude of this scene is indescribable. The great gorges, hewn through the solid rock are lighted here by the fire in heaven, and shaded there, by the wings of night : the echoes of the tearing cataract exert a magic spell upon us. Like goats around the candle of the Infinity we seem, as the eye g others its impressions of the surroundings.

The ride into the valley is full of surprises, as we swing from peak to peak, around the short curves, now facing "El Capitan," now nearing the foot of "Glacier Point," now passing "Cathedral Spires," now opposite "Half Dome" and "Washington's Column," now in view of the "Three Brothers," now along the foot of "Sentinel Rock," and now under the spray of "Bridal Veil Falls," which hangs, like a misty veil over the rocks, then tumbles over the boulders below, into the Merced river. The river is dark and deep, reflecting faithfully its magnificent surroundings.

We speed along the river's edge, until opposite the "Widow's Tears," so named for two reasons: one is, they only flow three months, the other is, because of the Bridal Veil. The Sentinel House is now reached, located in a picturesque spot beside the river, and commanding a fine view of "Sentinel Rock" in front, and "Yosemite Falls" at the rear of the house. We dash along, however, a mile or more to the Stoneman House, situated with "Half Dome" and "Washington's Column" as a background, and Yosemite Falls as a foreground. These falls make music while we lunch, and sing us to sleep afterward, while we rest both brain and body for the trails we are to take to all the wonders of this far-famed valley of grace and eternal beauty.

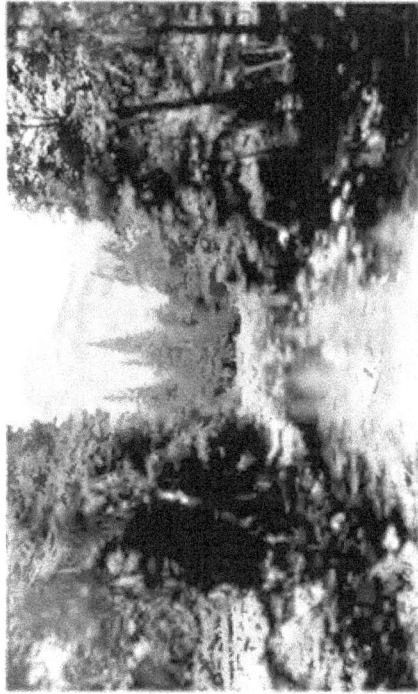

H, YOSEMITE! Yosemite's night rests like a benediction upon thee, and thy mornings are a psalm like prayer. Thy heart holds the secret of the Universe; and the key is withholden from man. Whether nature threw up these towering rocks in some awful spasm of effort to be freer and greater, or whether they are the work of a glacial period of time remote, we are left to conjecture. The evenness of the rocky heights, the gorges torn between them so smooth and clean, inclines one to the latter opinion. What but oceans of water in boiling torrents could thus wear away these solid rocks? What but water could *still* weep over the Jerusalem of its destruction and babble incessantly of its constructive genius as well, leaving as it has, these marvels of its handiwork for the enjoyment and instruction of a journeying world?

It rains upon the mountain tops today. Heavily the dark murky clouds hang there. While here it is light and cheery, though the air stirs the trees sufficiently to wake the echoes in the hollow glens, and cool the heated air. But hark! what is that peal on peal resounding through the valley? 'Tis a cloud burst on the mountains, that soon changes the white falls to a dingy yellow, and tumbles them down the mountains with renewed energy.

Somewhat like a pitched battle is a thunder storm in the valley. The clouds lower, and grow threatening; the valley darkens like a prison house, and as the wind begins to rise, the glens begin to whisper, softly at first then louder, the rain splashes in gusts, and now a flash of lightning reveals the great protecting rocky sides of our prison for an instant, then mutters its thunders against them. Soon the glens send back their reverberating answer.

As the fury of the storm ceases, the artillery of battle is appalling. The keen sharp report of thunder is carried along the valley and taken up by the glens, until seven distinct echoes, one following another, become a wonderful example of heaven's commanding. One almost expects to see destruction on every hand when the storm subsides, but no, out of the fury of these seven fiends, the tall spires begin to appear, until all are visible, more immortal than before. Their robes are cleaner and brighter, as they smile in the sunlight in proof that they enjoyed their little bath, and its melody as well.

CHILNAULNA FALLS—1000 FEET.

Just opposite our window is "Stair Case Falls," trickling down a rocky shelving or stair-like shape, which forms a part of Glacier Point. Along all these rocks is the impress of huge icicles, which hang from their edges from fifty to one hundred feet long, when the valley is locked in winter's embrace.

Looking from the porches of the Stoneman House, we are in view of "Glacier Point," "Eagle Peak," "Indian Canyon," "Cloud's Rest," "Royal Arches," and "Ten-ie-ya Canyon," also the foreground of falls and meadow and the background previously described. "Grizzly" and "Moran Point" are hard by, and just beyond are "Cathedral Spires," pointing like index fingers to the light above, as the ruling principle in this temple of the living God.

There are but two spires now, but the torn side of the rocky structure reveals the former position of another spire, which reared its towering head to the very skies. The hymns the waterfall is singing, the silent sermons in the stones, offer the divinest worship. All days sing praises, and all nights give God the glory, as century succeeds century, and nations rise and fall.

Taking a stroll in the morning air, we follow a trail to see where it leads us. Ere long we spy a bridge spanning the river, and jostled there are these words, "Happy Isles." We leave the trail and follow a path through the woods, soon coming in view of "The Rapids." A felled tree makes a bridge across them. We walk along to the end of the first island, across another bridge to where the rapids fork, and lo! a view of the rapids above us: boiling, seething, twisting, swishing along like the rapids of the Niagara, less voluminous, less boisterous, but recalling our trip to Niagara.

The name "Happy Isles" is no mystery now. Ha! ha! ha! laughs the water, as it tumbles over the great boulders. Jove has hurled down this tremendous chasm from the thrilling heights above, and ha! ha! ha! laughs back the answer from out the heart of the rocks and trees. Te! he! he! the river giggles as it hurries down the vale. Ho! ho! ho! the cataracts thunder, and the mountain tops reply as the wind sweeps down the chasm, bearing its message from the heights of "Liberty Cap," "Give us freedom or we die," white laughing down the gorge, breaking over the Happy Isles, filling our hearts with joy unspeakable, the cry is taken up and carried to earth's remotest bounds, "Freedom or we die." Unceasing vibration, the rest in activity, the "poetry of motion," the life of labor, was never better illustrated than here.

The heat from our rocky surroundings makes the valley a veritable bake oven from ten o'clock to three each day, so we write our descriptions of places and things, and sleep the heated hours away, when not on the trails enjoying higher altitudes.

" Mirror Lake " is our next prospective tour. We breakfast at six this morning, and at a quarter of seven step into the open coach that bears us through the fragrant pine woods, along the floor of the valley, beside the dusky Merced river, until the lake is reached at 7:15. This lake is an irregularly shaped body of water, skirted with trees and piles of mixed black and white rocks, resembling salt and pepper cloth. It lies in a " sweet sequestered spot," where the wild azalias grow, and seems a picture of living green at first, but as we make nearer approach we find that the water is clear as crystal.

Straying through one of the " Royal Arches " the reflection of sunlight from the rocks falls upon the water and lights up the scene, showing the topmost part of the mountains in the bottom of the lake, where the size of mountains, rocks and trees is magnified, and the colors, always beautiful are mysteriously intensified. The slight ripple of the water stirred by the beat of an insect's wing, or the movement of the fish beneath its surface, makes a changing, bewildering kaleidoscope of color.

Like the scenery of another planet one imagines Mount Watkins and Half Dome appear as we view them in their liquid grave, while Ahwahne (the watch dog) stands like the sentinel at the door of some new earth. The reflection of the skeleton limbs of the trees gives the scene a weird appearance, and suggests the fleshless forms of other lands than ours. The ferns, flowers, brush and rocks upon the water-edge, mirrored and magnified are beautiful beyond compare.

As we look, there gleams across the depths below, great arms of fan-shaped light, such as a search-light throws. We look above, and just behind, Ahwahne we behold the first fleet footed messenger, the sun is scudding to herald his approach, and thus the " Watch Dog " proclaims, " What of the night? " The trees on " Ahwahne's " heights now turn silver white as they tremble in the light, and soon the golden face of the king of day appears shining in the water. We hurry up the trail to the bridge to witness the second sunrise. Here the sun shows an outer rim of yellow light, the centre is black, afterward becoming silver, blue, gold, and finally red. We hasten on to a depression in the trail and view the third sunrise which tints the sky a glowing crimson from " Ahwahne " to " Half Dome." A rim of gold adorns " Ahwahne," the sun then turns to a brilliant emerald hue and sinks in the darkening waters, which have changed to a murky green, sombre and spiritless where a moment ago they portrayed a veritable palace of light. Altwahbe is suddenly transformed into a dark blue mountain, where anon it was as a flaming sword guarding the "Holy of Holies" that the sanctuary be not profaned.

P. S.   Among the visitors today was a woman who looked at the lake and its reflections awhile, then said: "I have heard people say that their first impressions of the valley were overpowering but to me, great rocks are just larger small rocks, and water is water the world over."   Fortunately or unfortunately for her, the writer failed to recall the words of Grace Ellery Channing's poem "Pity O God."   Otherwise down upon those rocks there would have been a kneeling form, and a prayer would have gone heavenward with all the fervency of the revival spirit, "'Pity thy blind O God,' and thy deaf and dumb as well, these helpless ones hemmed in by material sense, who never catch a whisper, or get a glimpse of the Eternal Spirit's manifested power; who having eyes see not, and ears hear not, neither do they understand."

A walk to Mirror Lake the day following the first visit there, reveals far more beauty in the reflected rocks than before.   The eye becoming accustomed, sees the magnificent distances of sky and land, more variety in the tinting of rocks, leaf and tree, deeper terraces in the mountains, phantom cascades and instead of three sunrises this morning there are five, each differing from the other.

Walking back across the bridge which spans the Merced river, the sun strikes the water revealing the river bed covered with golden brown boulders, varying in size from a billiard ball to a man's head.   The play of color is a study in brown and green, as the day advances toward the meridian.   There is no power of speech to tell the story that the lake, the river, the boulders and mountains tell us.   We look and gasp, and look again, and can only exclaim, "Mon Dieu! Mon Dieu!"

MIRROR LAKE.

OFF TO THE CASCADES.

UR NEXT trip is to the cascades. The afternoon is selected as the most suitable for this, if we reach the "Bridal Veil" in time to get the rainbow effect. Our way leads through much the same forest scene as before described, this time in the direction of Yosemite Falls. As we pass "Indian Canyon," the guide tells the story of the discovery of the valley. It seems that in time gone by the Indians were in the habit of stealing stock from the settlers west of the valley, and driving them into their own settlement. This resulted in their being followed and driven out through one of the canyons, which now bears the name of "Indian Canyon" from that occurrence. Only twenty Indians now inhabit this, their old home, civilization having taken from them the home the "Great Spirit" gave, and with it all the happiness life held in store for them.

We pass by the cabin of the old guardian of the valley. It is known as Hutching's cabin, situated in a picturesque spot, and is very attractive and inviting for so small a house. The spray from "Yosemite Falls" now begins to bathe our faces. The horses dash along enjoying it as well as we. The driver calls a halt, and we walk to the foot of the falls, which are three in number, and said to be the highest in the world, being about 2548 feet. The dash is tremendous as the water strikes the rocks below, baptizing everything in its reach. Near these falls we find the Mariposa lily looking so like a butterfly that the eye is deceived. These grow in a variety of shades, from pale cream color to dark heliotrope. The tourist can frequently be seen digging the roots to take home, that friends may see this "belle" of forest flowers.

We are now nearing "El Capitan." No wonder it is a "Great chief" to the Indians. So massive does it seem that one is lost in admiration and awe. It towers over 3000 feet above the valley, bearing upon its seamed, scarred surface, many a grotesque image, where each new corner discovers something more than his predecessors in the pictures Time has engraven there. It stands so nearly perpendicular, that we look straight up its sides. In one place there is a tree growing about half way up this piece of solid masonry,

EL CAPITAN

which looks like a mere speck, or outline against the rock, but the guide informs us that the tree is 120 feet high, 9 feet in circumference and 3 feet in diameter.

Hard by are the "Three Brothers," which together with "El Capitan," are mirrored below in the Merced river, which backs up into the fields and is as placid and true as a real mirror. Just beyond is "Eagle Peak," from which the guide tells us, may be seen the snow-capped Sierras, and a view of the valley looking Northeast. Southwest of us is "Buena Vista Peak," a well rounded, regularly shaped mountain of great beauty. Near by is "Pulpit Rock," so called because there lies across its top, a book-shaped rock, giving it quite a pulpit like appearance. "Leaning Tower" is now pointed out, also "Cathedral Rock" on opposite side of the valley. The top of Bridal Veil Falls lies between these two. Cascade Rock looms up ahead, and now the river is growing in interest every moment.

We are spinning along down a gradual incline ; the lower we go, the more momentum the river has, and the cascades, which began by tumbling and tossing over small stones, now dashes over great boulders 20 feet high. These are worn smooth by the water's action, and gleam through the snowy froth, like the eyes of some living creature. The echo of these cascades are like unto a grand anthem, rolling from ten thousand throats. As I listened, I no longer wondered at the old Indian's reply to the "pale face," who asked him after hearing a harp played for the first time, if the music was finer than any he had ever heard. The Indian shook his head and answered, "Much fine, but the Great Spirit's music is good enough for me." Just as one joins a chorus through sympathy, so we sang with nature a "Halleilujah chorus," spontaneously, as children sing when full of life and joy.

We now turn, and drive along the other side of the valley, close to the foot of "Cathedral Rock," and "Leaning Tower." These are sparsely covered with pine, fir, spruce and cedar trees, all standing, no one knows how, on the glazed and flinty rock, yet green as the grasses and moss of the marshes below. "Profile Rock" now appears and what a perfect profile! It is a man's bewhiskered face. The brow is the broad Grecian type, the nose the Jewish form, the eve is perfect and deeply set under the projecting brow. The lips are full and winning, rather than severe. Altogether it is a wonderful formation, and one can hardly believe that Time used only the wind and wave as mallet and chisel to sculpture so perfect a semblance to a human face.

BRIDAL VEIL FALLS.

What is that through the trees? That glint of strange light, now purple now gold as we go circling along? O! 'Tis the rainbow! the rainbow of the spray: We stop immediately before it to study its changing hues. High up in air, as the water from " Bridal Veil " splashes against the rocks, there is a rich orange color breaking away into green, violet, and blue as it appears and reappears like so many distinct though changing musical tones. The heart is so full of admiration, the brain so crowded with these rare beauties that we *must* relieve our pent-up feeling. So we shout approval as the rainbow dances in the spray like some sprite trained to color music. The great musician has now dropped his wand, so far as the falls are concerned, and we reluctantly turn our faces toward the hotel, to be confronted at the first turn in the road by a sunset view in the " Royal Arches," and upon " Clouds Rest," and thus ending the evening of the third day at Yosemite.

We are awake betimes next morning, for are we not going on the trail? Yes and ride a mule and wear divided skirts? Who wouldn't get up early to make ready for such new experiences as these? We breakfast in great excitement, and jump into the hack which is to carry the crowd to the foot of the trail, where horses and mules stand ready to bear their burden skyward; not "swiftly as on eagles' wings," but slowly, steadily, and persistently we afterward discovered. After seeing that all are properly mounted, the guide springs lightly into the saddle, starts some of the more experienced riders ahead, and allows the initiates to follow him.

The trail, none too wide to begin on, narrows as we ascend until in many places there is just space for the mules and their riders between the rocks and trees. The road which commenced in gentle curves becomes more and more zigzag, presenting the appearance of saw teeth ahead of us. Sometimes we are hugging close to the rocks, and again hanging over chasms, the depth of which increases as the distance grows. We soon reach the foot of " Vernal Falls," which tears down the mountain, the tallest, broadest, freest fall in the valley. On either side of the rapids that rage below it, the banks are soft and green, giving the fall its name and also color to the water below it. The rapids are like a seething sea, on which no ship can ride. The white dashing foam, the wild rolling tide fearless and unfettered plunges down, *down, DOWN,* into the boiling depths below, and the sight is as intimidating, as it is magnificent.

We dismount and walk a short distance to " Lady Franklin Rock," where we watch the flow of this mighty volume of water, which knows no ebb tide forever, but drives resistlessly on, roaring in its impetuous speed, and frothing like some hard ridden steed, bearing a message of mighty import to the slumbering world. "'Tis better on ahead," says some one, and away we scramble in high glee to remount, and continue the trail. No idea can be formed of the height of this fall, until one journeys up the side of it, to the top. The hard laborious task accomplished, we again dismount, and walk over mammoth "tables of stone," to the top of the same falls.

Leaning over nature's massive balustrade a look downward is appalling, the heart sinks, and the knees quiver, but the look repays. Lying there on the carpet of green, hundreds of feet below us, is a rainbow, true in color as any that ever arched the dome above. This "bow of promise" is seen both early and late, and is a sort of perpetual covenant with man. The force of the falling water is even better illustrated from this point than from below.

Just large enough to admit one person at a time is a crevice in these great beds of stone. It proves to be a cave, from the end of which one gets a view part way down the falls, showing how remote the water is from the rocks at that point. The estimate is, that it shoots out a distance of 120 feet or more, thus giving a still better idea of its dynamic qualities.

"Liberty Cap," "South Dome" and other gigantic grizzlies, tower still higher above us, and on beyond lies the falls, known as the "Nevada." One more look at the diamonds of the "Vernal," and of the cascades below, and we are off for dizzier heights and newer scenes.

In following the trail in the direction of "Nevada Falls," we cross a bridge over the rapids they form. Here the majesty of heaven and earth overcomes the brain. High above us, the white face and distant roll of the "Nevada" appeals to the eye and ear; below and behind us the roar of the "Vernal" continues, under foot the water is sweeping in a great silvery sheet (appropriately named the silver apron) over a rock smooth as polished marble. It is *one solid rock*, 85 feet wide and a *mile* long to all appearances. Above it the water is fretting, and snorting, and steaming along, when suddenly with a mighty gush it breaks, and sweeps over the incline in a large smooth volume of crystal splendor, at the rate of sixty miles an hour. Reaching the end, it bursts over the edge striking great boulders 30 feet across and 25 feet thick, howling in agonizing accents as it rushes onward. Straight to the conscience of us all, comes the command, "Lift the heart and bend the knee." Not a word is spoken. Nature "has the floor," as the vast panorama above, before, behind, below, is unrolled for our inspection. If our conceptions are not broader, if our convictions are not deeper, if our aims are not higher, the fault is all our own, for nature is no niggard in her revelations.

Homeward we begin the trying march. Here the test of nerve comes as we ride down the zigzag trail, at times almost perpendicular. We dare not cast a downward glance : "Up and forward keep the eyes," the kindly voice of the guide directs, and just as in the *real* march of progress, we look upward and forward, until the goal is reached. Oh, so weary, and yet happily, we trot back to our comfortable rooms, where a refreshing bath, clean linen, and a tempting dinner awaits us, which serves, together with a night's embrace in the arms of Morpheus to make us ready for another day's excitement.

ANOTHER OUTING.

⁂

UP HURRAH! Off again on the trail! This time to Glacier Point! Rain has cooled the air, leaving the meadows fresh and green, and freighting the whole atmosphere with odors of pine, spruce and fir, which is more tonic than a medicine, as we breathe it in great draughts and feel grateful that we live. The procession, as it wends its way up the mountain, has the appearance of a caravan upon the desert, and not unlike the swaying of the camel, are the sure-footed little mules as they trudge uncomplainingly along. The ascent is so steep that we are looking up to, when not looking down upon each other, from the zigzag road we are traveling, which becomes shorter and shorter each turn, until there is barely room for two mules upon each terrace. There are nine of these short zigzags in one place, forming the letter Z and resembling a flight of rickety looking stairs, notwithstanding the fact that they have a whole mountain for their foundation. From a little distance above and below, it seems that we could step from one stair to the other. Each succeeding stair reveals some new beauty. Now the green meadows and dusky waters of the valley below, now the nearer view of some cataract, now awful chasms over which our feet are hanging, and now the rocky walls of the opposite side of the valley.

One and one half hours of these interesting and startling changes, and we arrive at Union Point, not yet half way up the mountain. Here we dismount to rest both riders and mules. No sooner are the saddles loosened than the animals lie down in the sand and go to sleep, while the riders run about from bluff to ledge, shouting each discovery. We are 2700 feet above the valley. Gazing into its depths, we see passing teams looking no larger than dogs; men can scarcely be discerned. The river looks like a small creek, and the road like a narrow pathway. Homes seem flat and broad, while great trees resemble a young forest growth. Leaving "Union Point," we pass through acres of a stunted growth, known as "Chinkapin stubble." It's leaves are thick and shining, and its fruits encased in a burr similar to chestnut burrs.

We are now nearing "Sentinel Rock;" what a wonderful rock it is, standing 8122 feet above the level of the sea! "Surely the point we are bound for isn't higher than that?" says some one. "You'll think so," said the guide, "wait till we get there." As a matter of fact however, "Sentinel Rock" is 1110 feet higher we afterward learned.

"Agassiz Column" now rears its head above an awful abyss, as we swing around another curve. It stands 2800 feet above the valley on a mammoth rock pedestal, and is itself 85 feet high. Here and there across the trail there gushes streams of "living water," and here both mules and riders greedily drink. Such cool delicious water, more inspiring than the cups of Bacchus! Onward still onward, up greater steeps we slowly climb, to a point from which the glimpse of a house through the trees reveals the fact that the end of the trail is reached and causes us to hurry on and seek the cool shelter of the hotel veranda. We drop into the first available seat, shielding our aching eyes by covering them with our hands while we rest our weary bodies. It is impossible to see more for a brief spell; we feel dazed and color blind, and a sort of sea sick feeling causes us to wonder what the going down will seem like.

Finally we venture to take a look ahead. One look brings us to our feet, our very heart aflame, and every nerve aquiver. Spread before us is a scene so vast, that we gasp in its presence for something to say which will express our wonder and admiration. Chain upon chain of mountains, higher and yet a little higher than the one we stand upon, are pressing up against the sky as far as the eye can reach. In the foreground are "Vernal" and "Nevada Falls"; at our right, "Illilouette Falls"; to the left, like silver threads between the mountains, are the far distant "Chain Falls" stretching from peak to peak. Snow caps the distant mountains in folds like a robe of ermine. Clouds wrap the nearer ones in soft embrace. Back and beyond all, great dark banks are standing out in prophecy of an approaching storm.

"Better walk on a bit," says the cheery voice of the good hearted Irishman, who ministers to the travelers' wants, "a storm is comin'." We walk a few hundred rods to "Glacier Point;" a flag floats there, and under its milky folds, we stand looking again into the valley. Distance may lend enchantment in some cases, but the effect here is staggering. Sticks and rocks thrown down the mountain give no answer back. Horses and cattle look like crawling ants. People can be seen only by the aid of a field glass. The river is a thread of light. The roads are like strands of golden thread, tangled together. Roofs of houses seem to touch the ground. Haystacks look like ant hills, trees like mere points of green, and Mirror Lake like a small pond. Near us is a great shelf of rock called "Over hanging rock"; we climb cautiously upon it

THE THREE BROTHERS.

CATHEDRAL SPIRES.

as we venture a glance below, then beg to be excused, for the yawning gulf seems drawing us over, with an almost irresistible force. It is a relief to stand again on level ground.

"There are Washington lillies here," says the guide, "let's go and get some." We hasten along into the depth of the forest, gathering honey-suckles and columbines and inhaling the fragrance of the lillies which must abound somewhere near. Soon we spy them, and scramble about filling our hands with the delicate things, the like of which we never before have seen. Fiery eyes are gleaming yonder. What can it be? Half timidly we draw near. Is it a burning bush? It surely looks like one. We bend forward and gather the scarlet blossom of the snow plant, so radiant in its glowing garments, so perfect in its outline, so unlike anything else that grows that we, instinctively, link it to the story of the "burning bush."

Surging over the soul is the memory of that command of old, "Remove the shoes from thy feet for the ground on which thou standest is holy ground." As applicable today, as then, is this command, for all ground is holy that bears upon its bosom a symbol of the Creator's power, whether it is a rock, tree or blossom.

The raindrops begin to splash in our faces; we therefore beat a hasty retreat to the hotel, where a warm dinner is set before us, and while we eat we watch the clouds and listen to the din of a storm in the mountains. Like ships at sea the lighter clouds are tossed upon the darkening cloud billows, the lightning seems to almost cut the distant mountains in twain. The thunder, how it booms from peak to peak, gathering fury as it goes! Peal on peal, it strikes the rocks like blows upon an anvil, while the falling rain adds to the interest of the scene. There is a veil of water and thick black clouds between us and the snow-capped mountains. A light shower is falling here, but over there 'tis "Bedlam let loose." After two hours of storm and shadow, which has made the air uncomfortably cold, the taller mountains begin to rise out of the darkness and slowly the heavens seem "rolled back as a scroll" until all are visible. The cloud effects, as this is taking place, enrapture the vision and are rarely more beautiful the guide explains. The sun lavishly spreads every color in his royal old paint pot upon these scurrying clouds, exhausting all except the intense blue, which prevails here, and a stray streak of white to enliven the scene.

At 2:30 p. m. the Great Artist prepares his exhibit at the lower edge of "Vernal Falls," where the rainbow spreads over the ground for more than a mile following the misty spray. All colors play like lightning for about fifteen minutes, then the orange shades are lost, leaving the green and violet and blue. These, one by one, disappear while the blue alone lights the whole volume of water and reaches to "Nevada Falls" above, where at ten minutes of three the blue shades are plainly visible across its transparent body. Soon the violet, green, and orange tints are added. This stands like a land, or belt of seven hued lights against the falls. The water playing behind it, gives it the effect of a thing of *life* as well as light.

We cannot refrain from singing praises for this, the year of our jubilee. Are we not catching a glimpse of the "Celestial City" with its "gates of pearls" and diamonds, rubies and precious stones, and its "streets of gold?" Isn't that the "Cherubim and Seraphim?" Isn't "the river of life flowing in the midst of the city?" Isn't that the "Throne" over yonder, where the Day God is just laying down his silvery mantle for the night? Aren't those the white robes of the blest fluttering yonder? and isn't that the "harp of a thousand strings" in time and tune to the music of the spheres? Yes, it is all there. No gem is too precious, no metal too pure to adorn the palace of the soul, when it is fully awake to the beauty of its rightful heritage, it's legitimate heaven, and learns to appropriate, from the open book of nature, and the heart of humankind, the things that are its own. A little bald mountain stands out among the others quite prominently enough to cause the question. "What mountain is that?" "That is Mount Starr King," says the guide. "Rightly named," said we, for though a little one, he was "as a thousand" in the social, political and religious life of California. "Long live the memory of Starr King."

The day begins to wane. Tired of sight-seeing, we turn toward the land of shadows below us. The downward trail is one never to be forgotten. The lofty heights, the gapping chasms, the receding sun, the slow careful movement of our mules, the approaching valley, the sights that lie between, will be carried with us forever. When about half way down, the guide calls out "Halt," and "See the Star of Bethlehem." Far down in the valley, in a small basin of water surrounded almost entirely by trees, a ray of light shines, bright as the most brilliant star at night. "Why the star of Bethlehem," says one. "'Tis not an Eastern star?" "No," says another. "That is the star of Empire westward taking its way."

Gradually daylight fades into twilight, as we reach the floor of the valley. Night has in store for us a pyrotechnic display, far distancing anything we have ever before seen in that line. Bonfires of pine cones, and pine wood are built upon the edge of "Glacier Point." These light the heavens for a time, then the burning embers are pushed over the edge of the precipice. They come streaming down the mountain for a mile or more, in a cataract of flame. It is indeed a reign of fire for the half hour that it lasts, and under its witching spell, we seek the quiet of our room, to pen you this descriptive letter.

ONE DAY'S WORSHIP.

AWAKE at dawn this Sabbath day, for "'Tis wrong to doze holy time away" when this morning of all others, has been planned for the "Eagle Peak" trail. Eagerly we spring into the saddle and away we canter to worship at Nature's shrine, leaving the pious(?) on the hotel porches to gossip over our sinfulness. We soon decide that the proper name for our crowd is "R. G. Ingersoll's Sunday School procession," and with due solemnity we begin the ascent, which for steepness has thus far no parallel. There are 152 short zigzag turns in the first mile: these are in places but two and one-half feet wide, along the side of a perpendicular rock, 1600 to 2000 feet above us, and over vast chasms that deepen every moment as we rise to higher altitudes. We are riding up the side of "Yosemite Falls," which are near 2600 feet high. The profile of these falls is one of the most charming views in the valley.

Near the foot of the lower falls is a tree, of so artistic a formation, that it bears the name of "Erin's Harp." A short distance above is the "Pool," which is a great basin, worn in the rock by the water, and is a seething, boiling, bubbling cauldron, as the water swirls and swishes through it. The smooth brown rock at the foot of the second falls, is some two acres across, and looks tearfully up at us just now, for the volume of water passing over it is very small through July and August. High upon the rock, near the upper falls, is a picture wrought, called the "Baboon," and sure enough, there he is, violin in hand, fiddling away in keeping with the music of the water.

We at last reach the top of the upper falls, but the guide says "Press on, while it is yet the cool of the morning, we'll stop on our return." We reluctantly obey, and the panting mules move on, up steeps that might daunt the courage of a rocky mountain goat. A few more curves, and we reach the "Eagle Peak" meadows, and oh, the soothing rest of the scene! In the centre of this level floor, lie the still waters of the meadow lake, bearing upon their bosom, the golden water lily. The sweet-scented grass here abounds, and nodding their heads in lover fashion, are the honeysuckles, columbine, yellow and blue marguerites, blue

YOSEMITE FALLS

la kapu, and countless flowers we never before dreamed of, adding both harmony and beauty to the *mammoth painting*, for such it appears to be as we see it spread at our feet. The glad mules plunge into the marshy soil and greedily devour the fresh grasses, while the pupils of the Sunday School drink in the pure sweet air, cull the lovely blossoms, and "view the landscape o'er."

After a mile or more of this matchless meadow we reach a beautiful forest skirting the foot of "Eagle Peak." Here we dismount and spread our substantial lunch. As the hungry Irish woman said we "ate with avidity and our teeth," like so many half starved tramps. The inner man satisfied, we stretch our tired bodies under the shading trees, and sleep the "sleep of the just" for an hour or more.

The impatient stamp of the mules who, tired of sleeping are ready to journey on, warns us that "after day cometh night." So admonished, we ride away toward the jagged uneven points, that seem to stand against the sky above us. It is only three-quarters of a mile to the top, but it takes an hour to get there. The view from "Glacier Point" is appalling to the senses, but *this* view is both majestic and picturesque. We stand in the centre of chains of snowy mountains, surrounded by gulches and chasms, innumerable, while thousands of feet below, the sleeping valley lies.

The impression one gets of this underworld from these awful heights, is silence, sleep, death. Encompassing us like a great "cloud of witnesses," these eternal towering rocks are standing. Not a whisper reaches us from the yawning gulfs and bottomless chasms, which seem to possess the power to draw us downward into their destructive embrace. Through the great rifts in the rocks the sunlight plays. Around, behind, before, the sombre shadows fall. We file round and round this mighty circle, discovering new points of interest at every turn. We devoutly worship the Indian's "Great Spirit" for the thousandth time, since our entrance to the valley. Sadly we wave a good-bye to these helpful preachers, and slowly retrace our way to the edge of the meadows, where we rest another hour to avoid the heat of the afternoon.

Emerging from the meadow, we come upon some magnificent formations called "Castle Rocks," from their close imitation of the castles, found in the Old World. This castle stands over 3,000 feet above the valley, and here at its base we find the dead body of a mountain lion, which having wandered too near the castle wall, plunged down to its death, crushing its body to an almost shapeless mass, "a thing which seldom happens," says the guide.

Jogging along, we finally catch the sound of rushing waters; quickening our pace we soon reach the top of "Yosemite Falls," that is, the rock where we hitch our mules. The path to the top is not as easy as it looks. We travel a long distance, down rough stone steps, finally clinging along the side of the mountain to an iron railing imbedded therein, then down more steps and through narrow crevices, we push on until the top of the falls is reached. A great shout bursts from our throats spontaneously. Above what is known as "Upper Falls," are two short falls, neither of which are insignificant. Then there is a great smooth inclined rock, which the water gushes over to the top of the falls proper. Hollowed into this solid mass of stone is a perfect bath tub, filled about three-quarters full of water. This is both inviting and suggestive, as the S. S. procession begins to look badly demoralized.

On a point of great prominence above us, is a granite formation closely resembling the thumb of the right hand ; this is called "Hutchings'" thumb in honor of the old guardian of the valley.

We stand and view the spray, as the water dashes against the rocks with such force as to divide it into tiny shot, each shot a diamond, when the sun is not turning them into jasper, amethyst, and gold interspersing with turquoise and ruby. We watch these changing into diamonds again, and watching, listen to the glad triumphal song of this trinity of choirs, as its voice echos through the canyons and strikes the vaults of heaven.

One need not hold in anticipation the "song of redeemed souls," after hearing *this* sanctified song of nature and her creations which *has* rolled throughout the cycles of the past, and *will* roll down future ages. "The new song" it is singing *now*, to both the ransomed and the unregenerate throng, who have the opportunity to hear and the consciousness to appreciate the glorious flex of its mighty voice as it rises and swells forevermore.

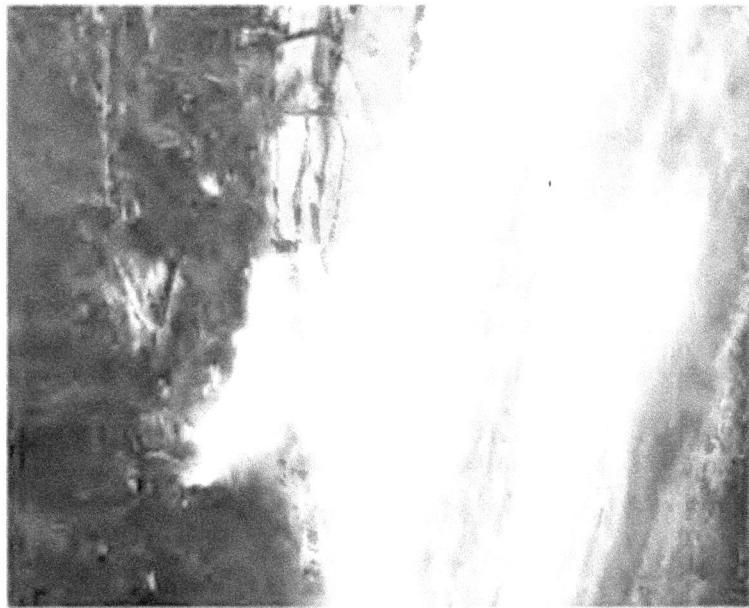

SILVER APRON CASCADE

VIEW FROM EAGLE PEAK

ON THE HEIGHTS.

SCARCELY has a new day arisen ere we have decided to attempt the longest of the trails which leads to "Cloud's Rest," the highest accessible mountain. We begin our preparation at five o'clock in the morning for this trip, as the mules arrive at 6.30 and we must breakfast first and see that the necessary luncheon is ready. It is seven o'clock before we are well started up the trail, which, leaving "Happy Isles" and its laughing water behind, leads up the Merced river for a mile or more in the direction of Vernal, Illilouette and Nevada Falls. The first place of interest is "Point Rea," at which place we suddenly come in view of the "Upper Rapids," and "Panorama Rock." We are riding along a shelving of rock at great distance above the rapids, from which we are shown the pictures, or the panorama. Here is a landscape, there a great bear: there are fish large and small, and there is a lady on horseback. One could sit for hours discovering new images in the worn sides of this flat surfaced flinty formation.

Soon we are riding under the shadow of "Register Rock," so called on account of the names painted upon it by travelers of all climes. Now we come in view of the "Nevada Falls," and slowly make our way up the steeps thereto. Only the "stairway" of the trail gives one any idea of the height of these falls. We are told that "Here men faint, and women have hysteria," and as we ride along the profile of the "Nevada," in the very face of its rushing water, heights above and depths below, we do not wonder much that people lose their heads and become childish in their weakness.

Dismounting, we walk to the top of this cataract. Far below us the "Silver Apron" is sweeping on to the top of the "Vernal," which falls over the straight even edge of its projecting shelf, as water might fall over a dam, constructed according to a well outlined symmetrical plan. Not so the "Nevada." It is the wildest of the falls, plunging over terrace after terrace of rock and leaping madly down to the rapids below. It roars and bellows and tears, as it speeds along. Great basins of various shapes and sizes, have been cut in the rocks about it, by the fury of its waves. It is the "Dare Devil" of the valley, and never ceases

NIAGARA FALLS

ON THE TRAIL TO NIAGARA FALLS

muttering, or splashing its spiteful tears.

From the mountain tops and down the serpentine river bed, the wind comes howling round the curve, over which the "Nevada" spends its power. "Liberty Cap," "South Dome" and "Grizzly Peak," in fitting dignity and uniform, stand guard over this reckless wanderer.

Before us in the dim distance is "Little Yosemite Valley." We do an immense amount of climbing before this destination is reached. It nestles so quietly against the tall mountains, that we think its real name should be "Peaceful Valley." Enjoyable beyond measure is the journey through the vale, with its blazing stars, lilies, cowslips, oxalis, and other native plants and flowers, together with it's murmuring river, skirted with trees and shrubs.

And now our wonderment begins, and constantly increases with each new scene. There are projecting rocks at the tops of ledges assuming shapes grotesque. Among them is a well defined alligator's head, a woman's face, cunning scrolls, a snake's head, and an umbrella spread over a queer little figure, an exact counterpart of a monkey. Here we get a view on a level with "Sentinel Dome" and "Glacier Point," bringing us to a realization of the heights we are climbing. We are squarely abreast of the snow-clad mountains, winding along through a low growth, bearing a white blossom resembling Scotch heather. We are informed that this is "snow-brush." Looking about us the name is no mystery, as the unaccustomed eye might easily mistake it for snow a short distance away.

Great pyramids are cleaving the sky miles above us. These are formed of layer upon layer of flat stone, piled in the utmost precision, making a well proportioned tower, a substantial monument to the "Unknown God." To the top of these pyramids seems a long weary trail for the already tired mules, as they slowly and cautiously ascend, passing the most dreadful looking "slides" of polished flinty rock we have yet seen. Their glazed and glaring surfaces are the most terrible record yet written in this great volume of surprises. At last the mules can go no higher. Leaving them to rest, we follow with hearts beating fast and faster for the lighter air quickens the pulse and almost stops the breath, as the guide leads the way to the top of one of these massive pillars or pyramids.

There sits a sphinx facing us, while around us grows the fragrant mountain oxalis. Here human hands have piled a monument of small stones, and ours make also such contribution to the memory of this marvelous journey and this matchless scene. Gathering courage as we stand on our lofty tower, in the rarest air we ever expect to breathe, we commence a survey of our surroundings. We are near 10,000 feet above the level of the sea, and looking toward the four corners of the earth, there is such a stretch of scene magnificent, that attempt to describe it is useless. As well stand at the Cliff House and try to describe Asia.

The valley we left this morning appears like a small green sward. "Little Yosemite Valley" is a 50 vara lot. Chains of mountains succeed chains as far as the eye can perceive, only one of them higher than the one upon which we stand. The blue waters of "Lake Ten-ie-ya" shimmer to the far north. Mirror Lake is a mere basin of water afar down the vale. Reverently we walk in spirit with Father Ryan, the sweet singer of the South, "Down the valley of silence, down the dim voiceless valley alone." In spite of ourselves we are stilled by the spirit of grandeur and greatness, until the thrill of the pulse of the Universe is felt and appreciated, as we are carried back to the time when the earth was new, and forward to where it shall be no more. Long and wistfully we gaze, for we well understand that such opportunity comes not again to most lives, and that ours are no exception to the rule.

"Come we must go or be belated," says the guide, and we are soon wending our way back to the valley and the plain. The hillsides below us as we are leaving "Cloud's Rest," look like tented fields, so white, so uniform and pointed are the rocks, scattered here and there in groups and lines and single places. The river has taken a dark green hue, as it rolls along beneath the gathering shadows of the woods. The "Devil's Slides" seem bound to precipitate us as we ride at a really safe distance, but seemingly dangerously near. Traveling again the floor of "Little Yosemite," brings a grateful relief from nervous strain.

Leaving "Nevada Falls," and its double rainbow again in the distance, as we make the difficult trail down its ragged sides, we reach ere long the top of "Vernal Falls." Here we leave the guide, who with the mules goes around the trail to the foot of the falls, while we climb down the ladders and take the footpath to meet them. These are wooden ladders leading under a sort of cave, formed by a large overhanging shelving of rock, where the five-fingered ferns and other kindred families flourish. At the foot of the stairs is a platform of solid masonry from which we watch the changing color of the waters, and the mossy carpet below. Winding down one flight of stone stairs after another and slipping as we go, for the steps are wet with spray, we reach the foot of the falls thoroughly drenched, and as we wander along we seem walking in fire mist, as the sun paints a gorgeous rainbow with his last lingering pencils of light. We seem for an instant "Floating in light with pearly gates near." The sun-shine kisses the river, and sweetly, tenderly leaves it for the night.

Clambering up "Lady Franklin Rock," we spy the pleasant face of the guide, with our faithful mules, as they swing round its curve to bear us home. Seated in the saddles, shouting a happy farewell to the sparkling falls, and their fountains full of jets of new meaning to our souls, we kiss our hands in loving good-bye to the river, and the crimson sky above it, making a picture that cannot die. No! never, since we are God's, and God is ours! Since nature is his body and humanity his soul, the picture will *live!* We shall find it again, as the "Brook is found in the river," the "Spent arrow in the oak," the "Song of love in the heart of a friend," and the fullness and beauty of all things, where the mind's polarizing power may call it into being.

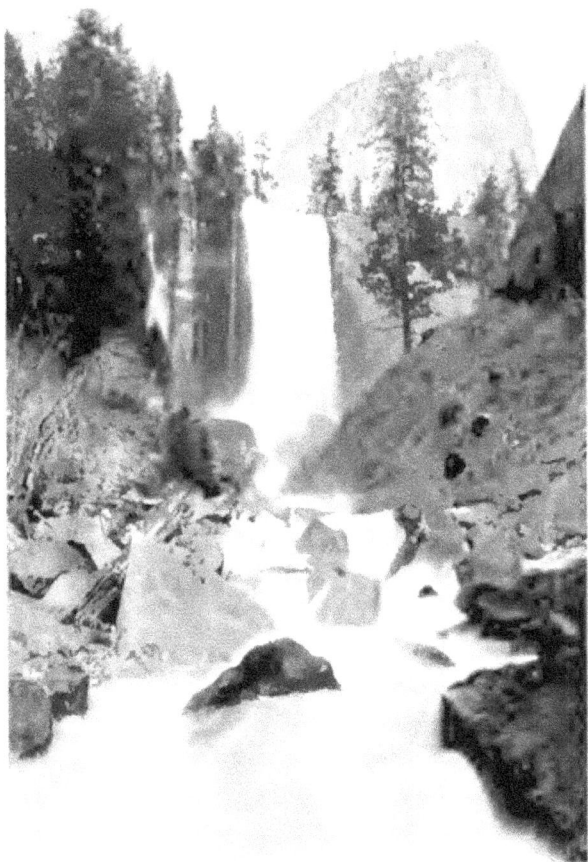

VERNAL FALLS AND NORTH DOME.

OUR LAST DAYS.

AVING made the last trail, we sigh for other fields to conquer. Accordingly we sally forth, lunch basket in hand, for a walk to Yosemite Falls; also for the purpose of exploring where the mules cannot climb. There are two miles of delightful walking, ere we rest near the foot of the falls, singing the while we rest and afterward clamber over great slippery rocks to the pool described in a previous letter. The spray dashes over us without mercy, baptizing us in the name of the All Mother. Thus blest, we leap from boulder to boulder, over the babbling, noisy water, drying our wet clothing in the sunshine, as our efforts increase with the distances we aim to span. Reaching a secluded spot, we bathe our heated feet before attempting the foot trail leading back of the falls. Cautiously we make our way, remembering that "Sinners stand on slippery places," and that care is needed lest we are lost in the abyss below. We are now clinging along the walls behind and under the roaring cataract, trying to peer through its mystic veil. Without success, however, as nothing but pure white light is revealed through the water. Drenched and chill, we gladly crawl out into the sunlight again to dry our dripping garments, which are soiled with dirt and sand from creeping rather than walking the last half of the distance up the slippery grade, glad to find the free air once more.

After an hour or more of the evaporating process, we travel on to an elevation above "Columbia Rock" for a last lingering look, as we have but one day more in this "Kingdom of Heaven." Descending we spread our lunch on "Columbia Rock," or rather a small portion of it, and sweeten our meal with plans for the outings we are to take on our next trip to Yosemite, for we have decided to rob a bank, or even economize in order to come again. We follow this by a noon day nap, continued until a light rainfall starts our party homeward. The rain ceasing before we reach the meadows, we spend the rest of the day gathering grasses, Mariposa lillies, and ferns, returning with hands and hearts full of the days attractions and pleasures.

GLACIER POINT.

"Go to Mirror Lake," says one, "Go to Happy Isles for a picnic," says another. "Let's do both," says a third. "Both it is," we say. Rising early we go to see the sun sail over "Ahwahne's" crest once more, and get a final view of "Mt. Watkins" and "Cloud's Rest" in the bottom of the lake.

Breakfasting we trudge off to "Happy Isles" where we write letters, gossip or sew as inclination dictates, until it is one o'clock before we realize it. We grow suddenly hungry, and lose no time in emptying our lunch boxes. The air is growing cooler, but we fail to take the hint until the grass is heavy with raindrops. We seek shelter in a thicket, thinking it is only a passing shower. But the rain pours, and the lightning's gleam, followed by peals of thunder, assures us that we are out for a rainy day. The ground being soft, we dig fern and azelia roots to transplant in our city gardens in memory of the day.

We are completely drenched and think it is high time to get back to the hotel. The boom of thunder, the almost incessant flash of lightning, the tumbling of rocks from Glacier and other points about us, together with the sheets of falling water, admonish us that we would better seek the thoroughfare, and keep in the "Middle of the road" in true Populist fashion. This we accordingly do, wading in water to our ankles in many places, but dragging our precious ferns along. A sorry spectacle we present as we sight the hotel, just as a carriage is leaving to gather us in from the highways, or rescue us from the raging streams.

The amused guests are of course assembled on the veranda, or standing at their windows enjoying our sorry predicament. We assure them that this is direct answer to prayer, as we have earnestly desired all the experiences of the valley. Between the porter and the maid, we are finally towel in and made presentable.

About half past four the storm breaks away and such a clearing we never again can hope to see. Clouds stretch like cobwebs from mountain to mountain, then smoke like, rise and sail away. The Western sun lights the fires in every raindrop attached to leaf or grass or tree, giving a soap bubble effect to the shining drops as they are shaken to the earth by the evening breeze.

An hour of this shifting scene and all is calm and peaceful, excepting the falls, which gush perfect torrents of yellow brown water, carrying great boulders and sometimes trees along, dashing them to powder or splinters upon the rocks below. The hurry of the water soon reduces the falls to something near their normal color, and at nine o'clock we are off to see the "Moonlight ride the bounding tide" of the beautiful silver cascades. Like the scenes of some fabled story are the views of Yosemite, when the "Curtains of night are pinned back by the stars, and the beautiful moon leaps the skies."

Every "Royal Arch," every flint like wall, every mountain top, every gurgling stream, and every streaming fall, is intensified in both size and beauty. The white light of the trees, as the moon's fairy fingers touch their branches, rising as it does five times each night, is a picture the God's might covet. The weird appearance of the rocks, flooded with silvery light is indescribably beautiful. The river, as it catches the silver kisses and buries them in its depths, is a veritable "Rogues' paradise."

The sheen of the falls beckons like an angel's wings, as it waves us its moonlight welcome. Phantom ships ride the white crests of the cascades and sail out of sight.

Streaming through the "Royal Arches" are great fans of translucent light, giving an etherial dreamy effect, enhanced in its beauty by the changing cloud drapery of the moon. Fantastic faces gaze upon us from every mountain top, for the moon is a wonderful sculptor, creating her imagery grotesque. The moonlight dancing on the lake is our farewell view of the valley by night.

It is morning now, and we visit the Indian camp, where we see the cunningest of Indian babies, and the mother Indians pounding acorns into flour from which to make their bread, which is neither an inviting or healthful looking mixture when finished. Aside from a few grass baskets which they have made for sale, this tribe shows no signs of energy or enterprise.

The one o'clock stage will bear us away from one of the most interesting spots on earth. When laboring midst the city's dirt and strife, or when the burdens of life chafe the spirit, 'twill be a joy to know that there is on earth a land so divine, that "The morning wakes in gladness, and the noon the joy prolongs, while the daylight dies in fragrance midst the burst of nature's songs," and knowing it, will be the lifting power over many a hard crossing in life, and the inspiration to future endeavor.

POST SCRIPT.

BEHIND us like a dream. lies the glorious Yosemite. Stretching away before our vision lie the mountains and meadows, the serpentine road and the river, each increasing in beauty every moment as we, homeward bound, rejoice in the changes the dying day is making. The while the soul of the day is passing into the realm of night, to wake again in the morning, we realize more fully the passing of human souls into the darkness of death to wake again in the morning of a new day. Nature is a unit, and because day merges into night and *vice versa*, our night shall merge into day throughout life's endless expressions. Musing thus, we arrive at Wawona just in time to enjoy the witching hush of twilight, which lengthens into night ere we reluctantly retire, knowing that the morrow bears us away from the marvelous beauties upon which the mind has feasted these fourteen happy days.

We set out at six o'clock next morning for the "Big Trees." Our way leads through the most stately forest we have yet entered. Banished from our Eden as we are, these sweet scented woods thrill our hearts with hope of a paradise regained. There are eight miles or more of this before we are rewarded by the object of our pursuit, the great "Sequoia Gigantea." Although we have heard and read of the enormous size of these fathers of the forest, we are unprepared for the sights awaiting us. Lying by the roadside are some of these monstrous creations slowly crumbling to decay.

So great in diameter is one of these tall and mighty skin. appropriately named "Dead Giants," that its projecting edges rise far above the stage coach and the heads of those sitting upon the box seat. Above us, 250 to 350 feet, these grand old giants stand, many of them wearing as breast-plates the names of such familiar cities as New York. Boston, and Philadelphia. Some wear the names of the States of Ohio, Michigan, Iowa and Illinois; and many more are honored by such titles as Abraham Lincoln, Col. Ingersoll, Gen. Grant, Longfellow, Whittier, Susan B. Anthony. Anna H. Shaw and other noted personages.

Chief among them all is "The Grizzly Giant," 285 feet high and 92 feet in circumference. We stretch our arms around it, to find that it takes 21 of our party to measure its girth. One of its limbs is six feet in diameter, and some learned professor has computed its age from the concentric growths, and declared the old veteran has stood the test of 4000 years. In the "Haverford," having girth of 96 feet, 16 horses have been known to stand at one time. Our coach loaded with passengers, easily glides through "Wawona," which is over 26 feet in diameter. The "Workshop" has a hollow at its base, measuring some 16 feet across. The guide states that there are just as many "Big Trees" in "Mariposa Grove" as there are days in the year.

The "Fallen Monarch" lies upon its side with its body in various stages of decay. We reach its top by means of step ladders, the one end requiring nine steps, the other eleven. Walking the length of this remains of former strength, we are once more reminded that "In the midst of life we are in death." Almost pathetic is the picture of this colossal ruin, with its shorn limbs, ragged bark and breathless body, lying midst more insignificant living members of its remarkable family.

The "Telescope," a big brother of "Old Grizzly" is broken off at the top and burned out in the inside. At the bottom is an arched doorway accommodatingly "etched" by the flames. Here we enter and "stage gaze" through this primitive though massive telescope, whose revelations are meagre in comparison with those of more modern invention, yet we are delighted with the lesson in the geography of the heavens, and pronounce it worth the time bestowed.

Another tree unnamed, its whole interior burned out for a distance of nearly 150 feet, lies upon the ground in the unbroken slumber of death save for the mountain stream which gurgles through it, enhancing the charm of this wondrous woody cave. Here stands a tree which has been broken off by some destructive storm, about 30 feet or more above the ground, in form resembling a broken column. Over its splintered edges a delicate vine is trailing, in its hollow top a variety of plants are growing, while over its sides the moss is clinging, making it as attractive as though some experienced florist had designed its decorations.

Countless other trees there are worthy of mention, but no description however adequate, can convey an idea of the magnitude of these magnificent sires of the forest, or the grandeur of their environments. We had expected to find superb pine cones in "Mariposa Grove," and were longing to reach the biggest

CATHEDRAL ROCK AND DISTANT VIEW OF BRIDAL VEIL FALLS.

www.ingramcontent.com/pod-product-compliance
Lightning Source LLC
Chambersburg PA
CBHW021428090426
42739CB00009B/1400